Robert THINGONTA

PASTURE PRODUCTION

Robert THINGONTA

PASTURE PRODUCTION

ScienciaScripts

Imprint

Cover image: www.ingimage.com

This book is a translation from the original published under ISBN 978-620-6-73484-0.

Publisher:
Sciencia Scripts
is a trademark of
Dodo Books Indian Ocean Ltd. and OmniScriptum S.R.L publishing group

120 High Road, East Finchley, London, N2 9ED, United Kingdom
Str. Armeneasca 28/1, office 1, Chisinau MD-2012, Republic of Moldova, Europe
Managing Directors: Ieva Konstantinova, Victoria Ursu
info@omniscriptum.com

Printed at: see last page
ISBN: 978-620-6-17195-9

STUDY OF THE DIFFERENT FACTORS INFLUENCING GRAZING PRODUCTION IN THE STUDY AREA: SPATIOTEMPORAL VARIABILITY AND DEMOGRAPHIC PRESSURE

TABLE OF CONTENTS

Summary

The spatial evolution of vegetation cover on the periphery of the PNSO is analyzed using three satellite images taken on separate dates over a 30-year period from 1990. The satellite image from 1990 to shows that the area of vegetation is 57722 ha, that of the degraded zone is 26178ha. And the area of agricultural land is 8619 ha. However, by 2022, the agricultural and bare land areas will have expanded by 32017 ha and 23593 ha respectively. And a decrease in vegetation area of 15462 ha.

Demographic pressure is diversified, according to our field surveys. In particular, 44% of the factors influencing grazing productivity come from rainfall, 34% from agriculture, and 21% and 1% respectively from pastoral activity and bushfire. Surveys of 148 households correspond to production units. Focusing on farm structures, the interviews focused on the general characteristics of production units, land use and land tenure and production systems. Classification according to habitat type, mobility and main activity enabled us to identify three types of farming: agro-breeder, agro-pastoralist and transhumant. Principal component analysis showed the existence of correlations between the six active variables used in this analysis. Four farm types were thus identified: pastoralists with livestock-only activity, agropastoralists with integrated activity (agriculture and livestock) but with dominant cattle rearing, agro-pastoralists practicing agriculture and livestock rearing but with dominant agriculture, farmers less equipped with at least one pair of oxen for harnessed cultivation, and transhumant farmers.

Keywords: variability, spatio-temporal, demographic pressure, actors, production.

1- Introduction

Anthropogenic activities through production systems are, along with climate, the main factors in the regressive dynamics of the natural environment. In areas of growing demographic pressure such as Mayo-Kébbi, extensive production systems are a real threat to wildlife, forest resources and soil quality. However, the various changes in agropastoral practices show that these production systems are undergoing radical change (Boutna, 2012). Studies on the evolution of herbaceous and woody vegetation in Chad indicate strong anthropogenic pressure on resources. Herbaceous and forest vegetation, for example, remain the main grazing reserves for an ever-increasing herd, leading to above-normal pastoral pressure. What's more, agricultural activities have sometimes developed on areas intended for pastoral activities, and consequently there is a significant risk of degradation of pastoral resources. The aim of this chapter is therefore to analyse the factors influencing the productivity of grazing land, in particular anthropogenic pressure on pastoral resources.

Materials and methods

The geographical coordinates of Sena Oura National Park and its surrounding area are limited: North-West between longitude 15°17'82" and latitude 9°18'69"; North-East between longitude 15°10'87" and latitude 9°08'07"; South-East between longitude 15°08'21" and latitude 9°07'52"; South-West between longitude 14°91'85" and latitude 9°00'04"; West between longitude 14°92'98" and longitude 8°98'56". It has 2,1300 inhabitants and covers an area of 1,502 km^2, i.e. 14.18 inhabitants/km^2 and 11.87 head of cattle/km^2. The area's population is growing at a rate of 3.5% per year.

2- Delimitation and space-time

Sena Oura National Park is located in south-western Chad, in the Mayo-Dallah department of the Mayo-Kébbi Ouest region, straddling the cantons of Dari (Lamé sub-prefecture) and Goumadji (Gagal sub-prefecture). It was created by law no. 11/PR/2010 of June 10, 2010, in accordance with the provisions of law no. 14/PR/2008 of June 12, 2008, governing forests, wildlife and halieutic resources. The surface area of the PNSO is (73520 ha) and that of the pérepherie is (150200 ha). The park is about 40 km from Pala, capital of the Mayo-Dallah department. The temporal delimitation consisted in situating in time a study concerning the pressure of agropastoral practices on the environment, as they are always imprinted by a vegetation dynamic in a reference time To (1990) and observation time T1 (2022). For this purpose, the subject is situated in a reference time interval To corresponding to the year 1990, and the observation time T1 corresponding to the year 2022. This time interval is justified by the possibility of obtaining recent information on the dynamics of the plant formation, since the study area is undergoing significant mutations due to human and natural actions.

3- Diachronic method

We used satellite images to determine the spatio-temporal factor from 1990 to 2022. Landsat imagery has a resolution of 30 m, so we were able to assess the evolution of ecosystem vegetation cover degradation in my study area. To do this, two types of classification were used: unsupervised ISO data classification to automatically group the closest classes, followed by supervised classification with maximum likelihood to determine the different land-use units.

4. Questionnaire survey and interviews

Survey forms were used to survey 148 households. The household survey covered two townships, namely Dari and Goumadji, located on the periphery of the PNSO. Sampling was based on household cards drawn up during the exhaustive census carried out in 2022. This census was then used as the sampling frame for the cross-sectional survey and for stratifying the sample. The questionnaire was drawn up in the form of an interview guide. The unit of observation for the survey was the farm or production actor (PA). In the 12 villages of two cantons, as in most terroirs in the Sudanian zone of Chad, a PA generally comprises a single household, that of the head of the family. Cross-sectional surveys focusing on farm structures were then conducted from November 7 to December 5, 2022. Ghiglione and Matalon (1998) consider these techniques to be more important than direct interviews for verifying information with the population. These surveys were carried out on the basis of sampling, which consisted in identifying a certain number of individuals in a population (sample) to whom biometric methods were applied in order to obtain a result representative of the population (Grawitz, 1990) cited by (Mamday, 2016). In this work, questionnaires on survey forms were administered to a randomly selected sample of 148 households comprising pastoralists, agropastoralists, agro-pastoralists and farmers, representing a population of 15,000 inhabitants. It is therefore a representative survey of the whole population, as emphasized by Balle(1992): "the *principle of a representative survey is based on the fact that a restricted population is studied, whose characteristics are presumed to be the same as those of the total population"* quoted by (Abdelkérim Youssouf, B., Korkorna, Bringue, M., 2016). The questionnaires concern the factors that influence grazing productivity in their locality. Given that the majority of the population at the site is illiterate, the questions were translated into local languages (*Moussey and Kado, Ngambay and Moundang*), and the forms were filled in as the respondent answered the various questions. Interviews were conducted with staff from public bodies responsible for natural resource management, heads of environmental departments and the ecological department of the Sena-Oura National Park.

5- Statistical data analysis

The data obtained were entered into Excel 2010. XLSTAT software was used for statistical analysis. Principal Component Analysis (PCA) (Philipeau, 1986) was used to analyze the information obtained from the surveys. In order to better characterize the combination of farms' main productions, six active variables were selected and analyzed: cotton, maize, sorghum and groundnut areas, number of draught cattle (traditionally an important discriminating factor among farmers), and the number of breeding cattle present on the farm at the time of the survey. This is the approach often used to characterize farm types on structural and technical-economic grounds. This approach is similar to methods for studying the diversity of production units based on a set of variables processed by means of multivariate analyses (Alary et al., 2000; Kobrich et al., 2003; Orsini et al., 1985). It better meets our objectives, as it is also based on a theory of the functioning of the systems analyzed, as already discussed by Landais (1996).

6- Spatial evolution of plant cover

The spatial evolution of vegetation cover on the periphery of the PNSO is analyzed using three satellite images taken at different times over the 30-year period from 1990 to 2023 (Figure 1 and 2). The satellite image from 1990 to shows that the area of vegetation is 57722 ha, that of the degraded zone is 26178ha. And the area of agricultural land is 8619 ha. However, by 2022, the agricultural and degraded areas will have expanded by 32017 ha and 23593 ha respectively, and the vegetation area will have decreased by 15462 ha. An analysis of the 1990 and 2022 topographical maps reveals a more or less considerable transformation of the natural landscape around the PNSO. Similarly, Akakpo (2006), sees this method as a means of helping to spatial changes. spatial changes. In addition, the diachronic study of land cover through a pseudo-directed hierarchical classification of three Landsat satellite images, enabled Wafo (2008) to carry out a multi-date analysis coupled with cartographic data and field surveys to highlight successive states of vegetation cover and to analyze its evolution between 1976 and 2001. Aoudou (2010) used a diachronic approach in his work, which, at the scale of the study area, covered a longer period of 50 years, comparing aerial photographs from 1954 and satellite images from 2004.

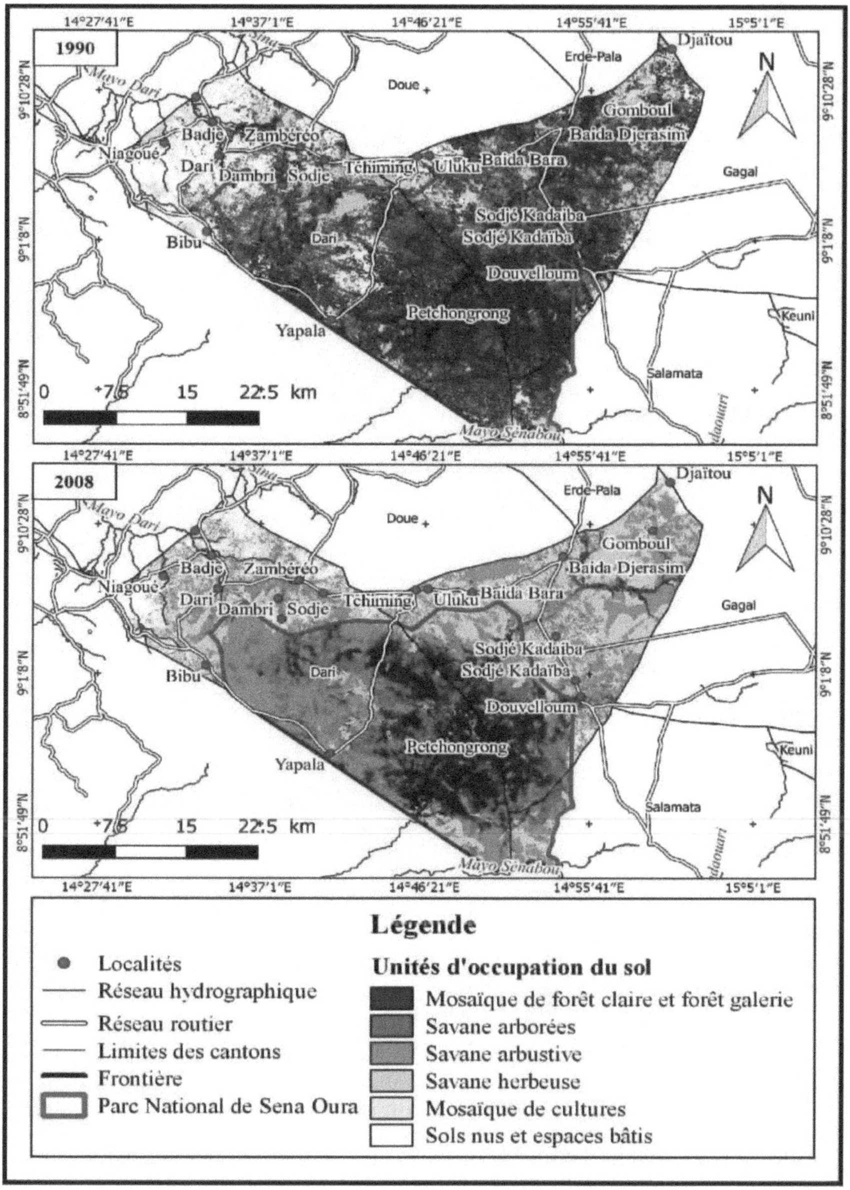

Figure 1: Diachronic analysis map 1990-2008

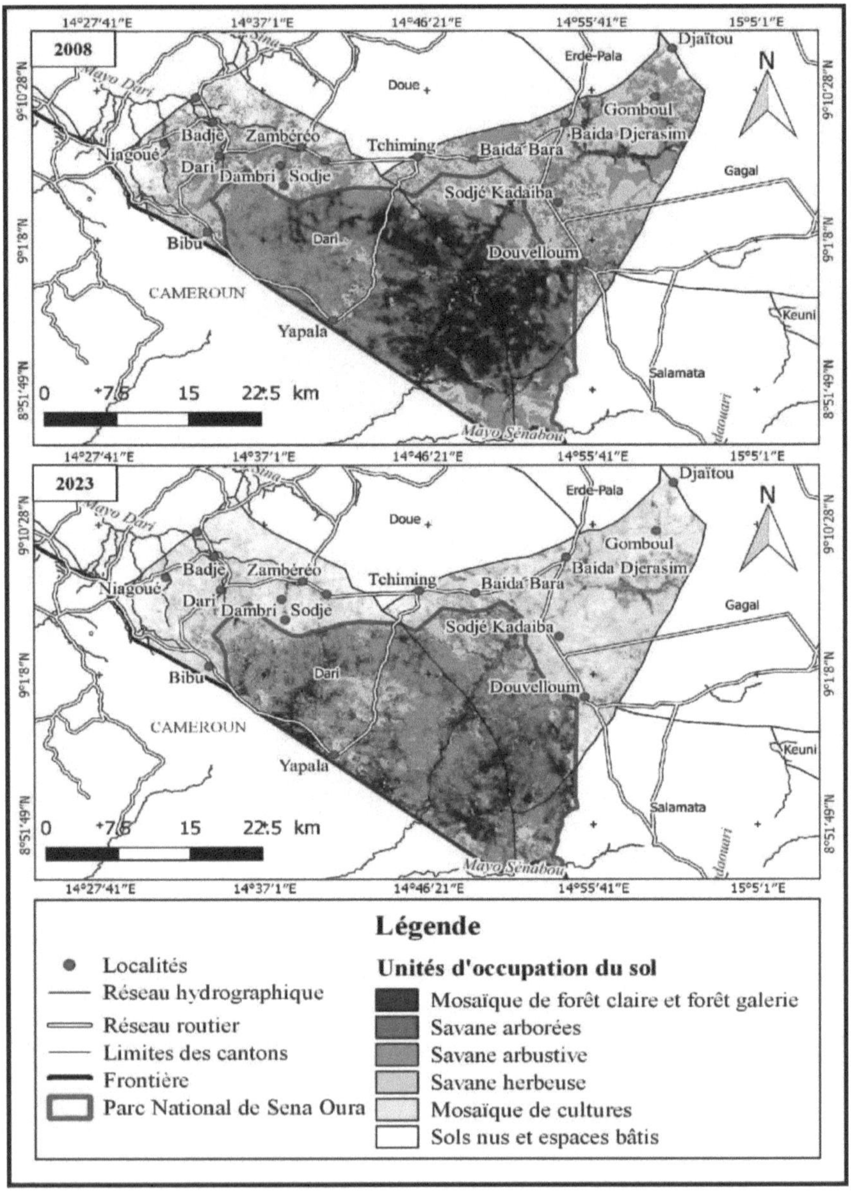

Figure 2: Diachronic map 2008-2023

7- Land use from 1990 to 2023

Satellite capture of land use shows, in descending order, the mosaic of open forest and gallery forest, tree savannahs, bare soil and built-up areas, shrub savannahs, crop mosaics and grassy savannahs. Bare soil and built-up areas, and cultivated areas are the units that contribute most to the dynamics of forest ecosystem occupation on the PNSO periphery. This land-use pattern translates significantly into a clear reduction in the surface area of natural formations in favor of anthropogenic formations over a ten-year period. The two satellite images from 2008 and 2023 show an increase in the surface area of agricultural zones, up by 2069 ha and 21329 ha respectively, putting pressure on the Sena Oura Park. Bare soil increased by 12990 ha, while grassy savannah was reduced by -11139 ha (table 1).

Table 1: Land use statistics

Land use units	Area (hectare)			Proportions (%)			Evolution	
	1990	2008	2023	1990	2008	2023	1990 2008	2008 2023
Mosaic of open forest and gallery forest	57722	23300	15462	42,42	17,12	11,36	-34422	-7838
Wooded savannah	27352	32089	15705	20,10	23,58	11,54	4737	-16384
Bare ground and built-up areas	26178	10603	23593	19,24	7,79	17,34	-15575	12990
Shrubby savannah	10420	28931	29973	7,66	21,26	22,03	18511	1042
A mosaic of cultures	8619	10688	32017	6,33	7,85	23,53	2069	21329
Grassy savannah	5789	30469	19330	4,25	22,39	14,20	24680	-11139
Total	136080	136080	136080	100	100	100	0	0

Analysis of this table shows that the 1990 land-use classes for open and gallery forest were 57722 ha, compared with 15462 ha in 2023. Wooded savannah covered an area of 27352 ha, compared with 15705 ha in 2023. Shrub savannah (10420 ha) increased by 29973 ha, which translates into abundant destruction of trees in favor of shrubs and grasses, which will occupy an area of 19330 ha in 2023. Cultivation had a surface area of 8619 ha in 1990, compared with 32017 ha in 2023. And finally, bare soil has increased by 12990 ha due to human overexploitation and natural effects. As a result, analyses show that the rate of increase in crop area between the two dates is 9.55%. The same observations were made by Gupta, 1971 "... demographic pressure has led to the cultivation of increasing areas of marginal land, withdrawn from livestock rearing; in the last 10 years, in Western Rajasthan, the cultivated area has increased by 49%, leading to a decrease in pastoral space of 15%; at the same time, livestock numbers have increased by 53.2%".

8- Impact of various factors on grazing productivity dynamics

Surveys of stakeholders revealed that the most recurrent factors affecting grazing productivity dynamics are rainfall, field extension, overgrazing and bushfire (Figure 3).

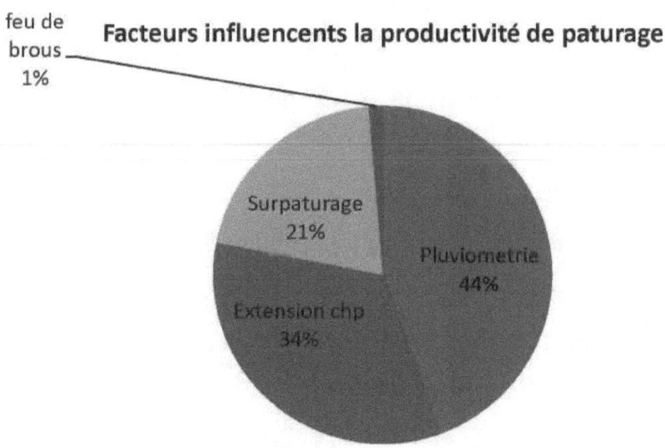

Figure 3: Factors influencing the dynamics of pasture productivity

9- Rainfall

The survey showed that 44% of the factors influencing grazing productivity come from rainfall. The diachronic analysis of rainfall in the PNSO and its periphery showed a drop in rainfall from 2000 to 2007, with over 1000 mm of rain recorded at the Galgal meteorological station located 25 km from the PNSO. Rainfall rose again from 2008 to 2014, with a peak of 1,397.46 mm, then fell from 2015 to 2021, with a slight increase in 2018 and 2019. The same observations on rainfall variation were made by Madjigoto in 2002 in the savannah of Chad: the average for the 40-50 decade was 1,252.9 mm, whereas it was only 902 mm for the 80-90 decade.

After a decline in rainfall in the 70s and 80s, there has been an upturn since the 90s.

10- Field extension

Surveys have shown that agriculture consumes 34% of land. The increase in agricultural production has resulted in the extension of rural space, leading to the shrinkage of vacant areas and increased deforestation. In the same vein, Abdoulaye, Bayi and Saidou in 2015 made the same observation about indiscriminate cultivation and restoration, abusive grazing and untimely bush fires, all of which are common human practices that lead to the degradation of ferruginous soils. The extension of cultivated plots in the village is encouraged by the cultivation of maize and cotton. Farmers continue to expand their agricultural areas. Their activities take the form of land clearing, stump grubbing and so on. These practices have an impact on the woody vegetation around the Sena-Oura National Park (Photo1).

Photo 1: field extension, (Tchingonta, 2023)

This photo shows the clearing of a new parcel of land for the cultivation of a field. This can be seen by the cutting or even stumping of ligneous trees. From the above, we can say that land clearing is a destructive factor for vegetation.

11- Shifting cultivation

Agriculture is the main activity in the villages on the outskirts of the PNSO, due to the fertility of the soil, the rainfall and the availability of arable land. It is practised in an archaic manner. The dominant crops are millet, white and red sorghum, maize, groundnuts, cowpeas, ground weights, sesame and cotton. The most dominant crops are maize, groundnuts and sesame. Following the cotton crisis, these have become the most important crops providing income for the population.

In this zone, the cultivation practices most in demand are slash-and-burn agriculture. These practices lead to soil impoverishment and the continual clearing of new agricultural land.

With the problems facing the cotton industry in Chad, agricultural inputs are becoming increasingly expensive. As a result, yields are falling, and farmers are forced to leave their old plots and burn other new plots to gain fertile land. As a result, the extension of agricultural land in this area is at the expense of the vegetation cover. To sum up, the plot management techniques most in demand by farmers living alongside the PNSO are slash-and-burn farming.

12- Pastoral activity in Goumadji, degrading practice of vegetation cover

In Mayo-Kébbi West in general and in the peripheral zone of the National Park in particular

In Sena-Oura in particular, livestock farming is diversified: diversity of species, diversity of farm types and diversity of pastoral practices. In this area, it is difficult to make a clear distinction between farmers and stockbreeders due to the different activities carried out by the farms. Although some claim to be farmers and others herders, agricultural and pastoral activities coexist (Boutna, 2012). The Sena Oura National Park and its surrounding areas offer a diversity of flora that is highly beneficial and essential for livestock feed. As a result, pastoral activities have a significant impact on vegetation cover, accounting for 21% according to field surveys. Given the severe shortage of fodder, young plants in particular suffer from recurrent trampling and grazing by animals grazing on the outskirts of the PNSO. There are also illegal activities such as pruning, severe pruning and abusive cutting of wood by herders to feed their livestock or for their own housing. Pastoral practices increase the degradation of vegetation cover, especially fodder trees. These factors are overgrazing, transhumance and bush fires to regrow fodder.

13- Transhumance

Transhumance is the mass movement of livestock in search of better pasture and water. In Chad, there are several types of livestock movement, although the major characteristic of livestock farming is transhumance (Béchir, 2010). In search of fodder and water, transhumant herders use a variety of pastoral practices, based on mobility and adaptation of daily rhythms to the availability of food and water. In this peripheral zone of the Sena-Oura National Park, transhumants stay from

October to April. During their mobility, these nomads severely prune woody plants for livestock feed (Photo 2).

Photo 2: Nomadic herder pruning trees in the National Park

Sena-Oura(Zoulinné, 2023)

In this photo, we see transhumants engaged in illicit logging activities to feed their livestock. The species pruned here is combretum reticulata. From all the above, transhumance is the main factor of pressure on the vegetation cover. The periphery of the PNSO has a variety of pastoral resources. This zone is coveted by breeders and farmers of all stripes. Allochthonous herders (Foulbés and Arabs) but sedentarized stay for a few months (June-October), i.e. until the end of the rains. At the end of the rains, however, the post-cultivation rangelands are gradually replaced by woody pastures. In fact, the availability of pasture in the area leads many herders to park their flocks in the same saturated area for long periods.

14- Bush fires, destructive use of vegetation cover

The use of bush fires is a simple and less costly practice. According to our field surveys, they account for 1% of environmental degradation. According to these herders, bushfire favours the regrowth of young plants (Boutrais, 1983), believing that bushfire is a normal and indispensable ecological factor in the maintenance of Sudanian savannahs. Grandi (1996), quoted by Mongo and Ngouné (2015), believes that fires generally leave bare soil with no vegetation cover in the Sudano-Sahelian zone. However, it should be noted that pastoral practices based on the use of bushfire are one of the factors that destroy vegetation cover.

Photo 3: bushfire as a factor in degradation (Tchingonta Robert 2023)

In this photo, the use of fire enables livestock farmers to obtain fodder resources. Here, a farmer is practicing this technique in the middle of the forest.

15- Production players in Goumadji and Dari canton

The classification of breeders is an old debate between several disciplines. It can be based on a producer's membership of a social group according to his or her historical origin: a Fulani farmer with just two head of cattle and several hectares of fields is still considered a herder. Similarly, a farmer with a hundred head of cattle is still a farmer. In the terroir of two cantons, the distinction between breeder and farmer is not obvious. In fact, it is exceptional to find herders who do not cultivate fields, or farmers who do not own a few head of cattle. In Central Africa, all stakeholders in production units built around the importance of livestock and its integration into the cropping system (Vall et *al.* 2003; Djamen et *al*, 2003; Jamin et *al.* 2007) have identified three main types of stakeholder: farmers, agro-pastoralists and traditional herders (pastoralists, Fulani, Showa Arabs, Toucouleurs, etc.). However, a comparative analysis of these different types shows that they do not always refer to the same empirical referents (Jamin et *al., 2007).* However, in the case of our study, we are also focusing on habitat-related criteria (Photo 4 and 5) filmed in the village of Sodje canton.

Photo 4: village sodje farmer's dwelling

Photo 5; Fulani village of Sodje

16- Agro-breeders

In both Central and West Africa, the term "agro-pastoralist" refers to a farmer who, in addition to draught oxen, has a large herd of cattle (over 10 head) and a significant cultivated area (Dongmo, 2009; Duras, 2006). These are sedentary farmers (Zimé N'Gambaye Moundang, Mussey Toupouris, Massa) who, thanks to the development of cash crops (cotton, maize and groundnuts), have accumulated money by saving in livestock. They are professional farmers with a fixed habitat. Their farming system is characterized by small-scale livestock and cattle rearing. However, two types of cattle rearing can be distinguished in this system: on the one hand, draught cattle rearing, generally consisting of one or two pairs of males per farm, and on the other, professional rearing based on building up the herd by purchasing breeding females. The main production objective is to keep livestock as a savings account, with livestock forming part of a supplementary economy that often corresponds to a degree of enrichment for the agricultural producer.

16 - Agro-pastoralists

These are fixed-point mobile breeders. They are distinguished from the first group by the combination of two habitats: one fixed and the other mobile. One part of the family stays in a fixed camp, with a small herd that forms a dairy nucleus, and the other part transhomes with the cattle for a period of the year. The fixed habitat (banco house or straw hut), embodies the practice of agricultural activities and the other, a tent made of matting or tarpaulin (Photo 5), refers to mobility. This group is made up of Foulata, Mbororo and Haoussa Peuls, who share the same pastoral space in both wet and dry seasons. There is an abundance of literature on them (Bonfiglioli, 1988; Petit, 2000; Sougnabé, 2010). They are described as professional herders, with livestock rearing as their primary source of income. To limit the sale of livestock and to cover their own needs, they farm small areas with food crops. Thanks to the judicious use of animal manure, this type of farming is recognized as one of the most efficient. The main production objective is to increase herd size through reproduction. Agro-pastoralists are in a more complete agropastoral configuration, with land marked by cultivation, a situation also observed among herders in the semi-arid steppes of the Maghreb (Bourbouze, 2000).

17- Transhumants

Their main characteristics are mobility and flexibility. They are the Arabs from the center of the country (Dourbali and Melfi), the Foulata Peuls from the Chari-Baguirmi and the M'Bororo Peuhls from Nigeria. They are often transients, crossing the Yamba Berté Classified Forest on their way to other territories. They may cross the national border into Cameroon or the Central African Republic.

18- Farm classification using Principal Component Analysis (PCA) methods

The first factorial design is represented by axes 1 and 2, and accounts for 100% of the variability (Figure 21). Their eigenvalues are 53.04% and 46.96% respectively. Farms vary according to area sown, family size, number of farm workers, crop rotation and number of draught and breeding cattle. Automatic classification identified 4 farm types similar to the results of (Sougnabe, 2010). While Awa et *al.* (2004) described 6 types of herders in a terroir in the same zone, studied by Ali in 2017, which they then grouped into three main categories.

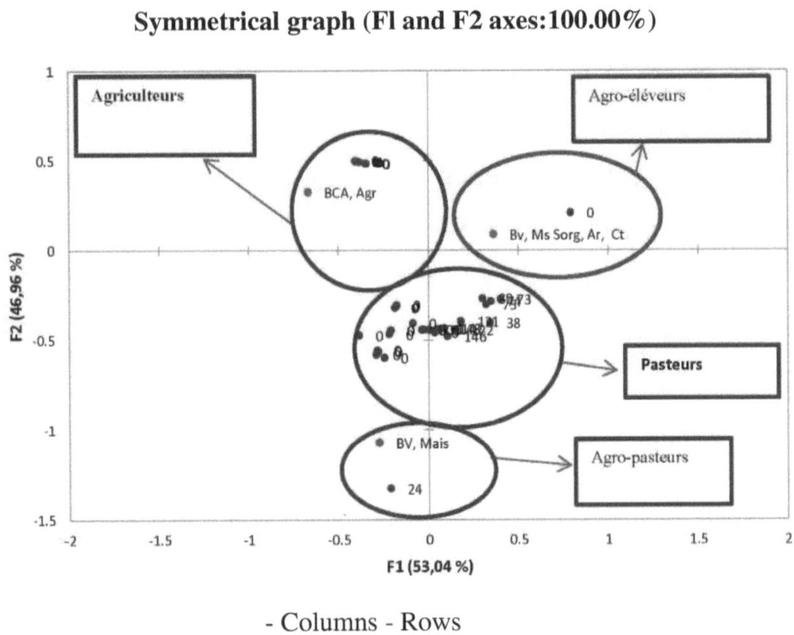

Figure 4: Principal component analysis

19- Type: Transhumant herders

This type of actor is made up solely of transhumants and represents only 6%. They come to stay in the locality for 4 to 5 months to take advantage of crop residues in the dry season. Most arrive in November and leave at the start of the rainy season. Their herds vary from 100 to 1000 head. The herd consists of that of the head of the household, his wife and children. They also raise small ruminants, asines and horses for family transport. These are Arabs from the center of the country (Dourbali and Melfi), Foulata Peuls from Chari-Baguirmi, M'Bororo Peuhls from Nigeria and Haoussa and Arab herders from Pala in the center of the Mayo-Kebbi West province.

20- Type AP: agro-pastoralists with predominant cattle rearing

This type is characterized by settled former Fulani and Arab pastoralists. It accounts for only % of total production units. Farm managers are 43 years old on average. Production units have an average of 6 members with 2 active workers. They grow maize and sorghum crops, and own the largest herds of cattle in the village (over 100 head on average). They also raise small ruminants (sheep and goats), which serve as savings for the family. The main cereal (maize) is produced not only for self-consumption, but also for marketing.

These agro-pastoralists have no land reserves. In the absence of fallow land, they must use organic fertilizers to maintain the fertility of the land. To do this, they use dry-season parking and the spreading of poudrette. They have the lowest cultivated area/asset ratio (0.2 ha/asset), compared with 0.4 ha/asset found by (Ali, 2010), as a significant number of the farm's assets are devoted solely to livestock farming and call on outside labor for field work.

21- Type AE: Well-equipped agro-pastoralists, farming and raising livestock

Agro-pastoralists of this type represent 27% of the production units surveyed, and cultivate almost all crops. They generally raise between 25 and over 100 head of cattle. Farm managers have the same average age (42). They have a sizeable land bank (20 ha), almost all of which is sown (94%). However, they manage to farm such an area without too much difficulty, thanks to a large workforce and a fairly high level of agricultural equipment. Agro-pastoralists have larger production units than farmers, with 6 workers cultivating around 15 ha to feed 15 dependents. The main cereals (sorghum, maize and cotton) are produced not only for self-consumption, but also for marketing. They sell most of their cereal production and almost all of their groundnut and sesame production. These traditional farmers have also become livestock breeders, capitalizing on their agricultural surpluses and sometimes extra-agricultural income in the form of livestock. They own their herds outright, and do not practice "confiage" (Ali, 2010). On the other hand, the number of small ruminants is modest (8 to 20 head). Some also raise pigs. Suckler cattle, draught oxen and other cattle are driven together to pasture all year round. In the growing season, the plough oxen are kept close to the cultivated plots.

22- *Type A: Medium-sized farmers, equipped with at least one pair of oxen.*

This type of farmer represents 36% of production units, and is characterized by the oldest farm managers on average (40 years old). They have a large land bank made up of cultivated fields, young fallow land and wasteland. They have a pair of oxen and a complete cultivation chain, but very few have a cart. Others, on the other hand, do not have oxen, but use herbicides instead. For these farmers, it is mainly draught cattle that make up the herd, unlike the other farm types (AE and AP), which have between five and ten head of dairy cattle.

Thanks to an adequate level of equipment, the cultivated area per worker is high (on average 2 to 3 ha). Their potential for fertilization through animal manure is not negligible, with a ratio of slightly less than 1 LU/ha cultivated, but remains low. They bring in a little yard powder and chemical fertilizer on certain plots (cotton and maize). Most of their production (maize, sorghum) is sold. Livestock production is limited to draught cattle, small ruminants and pigs.

Shifting cultivation

Agriculture is the main activity in the area, due to soil fertility, rainfall and availability of arable land. It is practiced in an archaic manner. The dominant crops are millet, white and red sorghum, maize, cowpeas, ground weights, sesame and cotton. During our surveys, it became clear that groundnuts and maize are the most important crops providing income for the population following the cotton crisis. In this zone, the most popular cultivation practices are slash-and-burn agriculture. These practices lead to soil impoverishment and the continual clearing of new agricultural land. With the problems facing the cotton sector in Chad, agricultural inputs are becoming increasingly expensive. As a result, yields are falling, and farmers are forced to leave old plots and burn new ones to gain fertile land. As a result, the expansion of agricultural land in this area is to the detriment of the vegetation. Of all the techniques for developing plots, the most popular with Goumadji and Dari farmers is slash-and-burn agriculture.

23- Clearing woody vegetation

Land clearing is one of the techniques, after slash-and-burn, that degrade the vegetation cover in the peripheral zone of the Sena-Oura National Park. Clearing is due to the high demand for agricultural space for cash crops (maize, cotton, groundnuts) and food crops (millet, maize, cowpeas, etc.). This small region, where a protected area is located, is undergoing major change, as it is experiencing a large spontaneous migration flow accompanied by significant land clearing. It is subject to strong anthropic pressure, and many issues are at stake: environmental, social, agricultural, pastoral, etc. (Sougnabé, 2010).

As a result of the above, the strong population growth has led farmers to clear large areas for cultivation from outside to inside the Park, despite the measures taken by the conservation service. This can be seen in the transition zone from Goumadji to inside the protected area.

24- Fire-sweeping: A destructive practice for woody vegetation

According to the Dictionnaire de l'agronomie, ecobuage is the clearing of undergrowth by fire. It's a technique that involves stripping vegetation and the surface layer of humus, incinerating these elements in small piles, then spreading the ashes on the land to enrich it with nutrients. Ecobuage is carried out progressively and, above all, at the start of field development, i.e. at the beginning of the rainy season (Abdoulaye et *al.*, 2015). It is well used in the peripheral zone of the Sena-Oura National Park.

During our surveys of farmers, the technique of "écobuage" (burning) was used to destroy vegetation. To this end, farmers burn part or all of a tree in order to collect the ash and spread it over their plots as fertilizer.

25- Plot management techniques on the outskirts of the PNSO

The percentages of landscaping techniques in the periphery are shown in (fig.5) below.

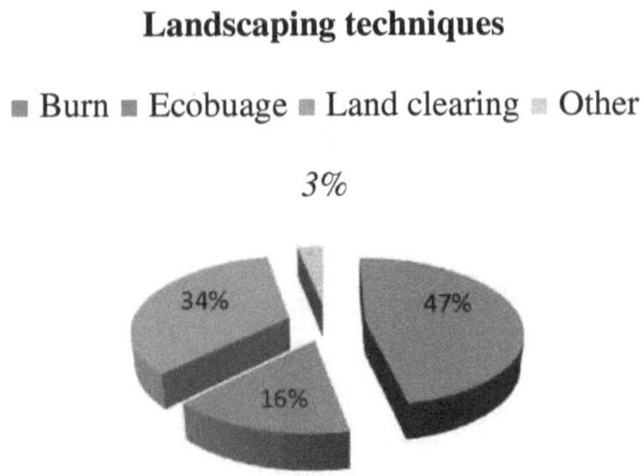

Figure 5: Landscaping techniques. Field survey, March 2023

Figure 5. Percentage of plots developed

Figure 5 allows us to analyze the different management methods used in the peripheral zone of the Sena-Oura National Park, specifically in the Goumadji terroir. We can see that the percentage of slash-and-burn methods is the highest at 47%, followed by land clearing at 34%, slash-and-burn at 16% and other practices at just 3%. As a result, farmers are turning to slash-and-burn cultivation, which enables them to extend their cultivation area while benefiting from low-cost ash fertilization. However, it should be pointed out that this slash-and-burn technique puts considerable pressure on the plant cover.

For data collection in the field, we used interview guides with more open than closed questions. Semi-directive interviews were preferred for individual rather than group meetings, to enable our interviewees to express themselves more freely and anonymously, without any form of pressure.

We have taken into account the administrative divisions in terms of cantons and villages in the vicinity of protected areas, in order to have a more representative sample of local communities. Villages are groups of permanent dwellings in the countryside (Larousse, 2010). In Chad, they represent the smallest administrative unit, with the exception of field huts and *ferricks* belonging to transhumant herders who settle in the locality for a short period or permanently. The huts and *ferricks* depend on the nearest village. A group of villages forms a canton, headed by a canton chief. The canton chief reports to the village chiefs. The stakeholders interviewed include nomadic herders, transhumant herders, agro-pastoralists and less-equipped farmers. The opinions of notables, in particular canton chiefs, initiation chiefs, land chiefs, etc., are of capital importance, as they enable us to appreciate power plays and traditional local governance. The focus was on understanding the dynamics of fodder production in natural pastures on the periphery of the PNSO, their perception of animal flows in the grazing area, and their level of involvement in the management of grazing areas using various participatory approaches to avoid overgrazing.

We have drawn up a list of twelve villages identified for the study, located on the outskirts of protected areas for survey purposes.

At the outset, discussions were held with the canton chiefs of Dari and Goumadji. Two 900 m^2 plots were fenced off on a selected rangeland and supplemented in the second year by 28 un-fenced plots to obtain a reasonable result. The data were collected in five main stages:

- An inventory of herbaceous and woody biomass was carried out on two fenced plots and 28 unfenced plots, covering a total area of 27,000 m^2;
- Assessment of Dry Matter Productivity (DMP) over a 20-year period: 2001 to 2021. The amount of dry matter (or above-ground phytomass production, DMP) in the rainy season is calculated from the cumulative rainy season decades (June to October): *PPA=0.01^LDMPdecadefdecade i* where PPA (kg MS.ha- 1) = above-ground phytomass production; 0.01 = conversion value in kg; decade i = first decade of June; decade f = third decade of October. The result is an estimate of total above-ground production for the year in question, expressed in kg MS.ha- 1;
- Survey of animal numbers;
- Factors influencing forage resources;
- The different resiliencies of pastoralists or agropastoralists in the face of dwindling forage resources.

Twelve villages were chosen for the household survey and for reasons of livestock concentration. In addition, focus group and individual interviews were organized to collect qualitative data.

The study of changes in forage availability was carried out over a series of years from 1991 to 2021.

26- Pastoral activity on the outskirts of the PNSO

Degrading the landscape

In Mayo-Kébbi in general, and in the peripheral zone of the Sena-Oura National Park in particular, livestock farming is diversified: diversity of species, diversity of farm types and diversity of pastoral practices. In this area, it is difficult to make a clear distinction between farmers and stockbreeders due to the different activities carried out by the farms. Although some claim to be farmers and others herders, agricultural and pastoral activities coexist (Boutna, 2012). The Sena Oura National Park and its surrounding areas offer a diversity of flora that is highly beneficial and essential for livestock feed. As a result, pastoral activities are a burden on vegetation in general and herbaceous vegetation in particular. Given the severe shortage of fodder, young plants in particular suffer from recurrent trampling and grazing by the animals of the pastoralists who graze them in the Park. Illegal activities such as pruning, severe pruning and the abusive cutting of wood by herders to feed their livestock or their own dwellings should also be noted. Pastoral practices increase the degradation of vegetation cover, especially fodder trees. These factors are overgrazing, transhumance and bush fires to regrow fodder. Our surveys show that not all species around the park can be consumed by livestock. They vary according to the animal species. We present some of the species eaten by livestock on the outskirts of the PNSO in the following tables.

Inventory of herbaceous species in the 30 plots Covers Height

Species	100%	(m)	Phenology	Family
Hypparrhenia rufa	5	2,3	fe3fr3	Poaceae
Schizanchyrium sanguineum	5	2,5	fe3fl3fl3	Poaceae
Crotalaria ononoides	2	0,7	fe3fr1	Fabaceae
Leptochloa caerulescens	2	1,2	fe3fl3fl1	Poaceae
Lepidagathis lutescens	+	1,2	fe3fl3fl1	Acanthaceae
Alternanthera sessilis	+	0,6	fe3fl3fr2	Amaranthaceae
Pandiaka involucrata	+	0,6	fe3	Amaranthaceae
Aneilema sinicum	+	0,3	fe3fr3	Commelinaceae
Cyanotis lanata	+	0,2	fe3fr3	Commelinaceae
Floscopalei thyrsa	+	0,2	fe3fr3	Commelinaceae
Ipomoea sp	+	0,3	fe3fr3	Convolvulaceae
Cassia mimosoides	+	0,2	fe3fr3	Fabaceae
Indigofera braecteolata	+	0,4	fe3fr3	Fabaceae
Indigofera colutea	+	0,5	fe3fr3	Fabaceae
Indigofera spicata	+	0,3	fe3fr3	Fabaceae
Tephrosia bracteolata	+	0,6	fe3fr2	Fabaceae
Hyptis sp	+	0,8	fe3	Lamiaceae
Hibiscus aspera	+	1,1	fe3fr1	Malvaceae
Melochia corchorifolia	+	0,5	fe3fr2	Malvaceae
Urena lobata	+	0,5	fe3fr2	Malvaceae
Andropogon tectorum	+	0,9	fe3fr3	Poaceae
Hypparrhenia subplumosa	+	1,5	fe3fr3	Poaceae
Loudetia togoensis	+	1	fe3fl3fr1	Poaceae
Panicum maximum	+	1,8	fe3fr3	Poaceae
Panicum pensum	+	1,6	fe3fr3	Poaceae

Pennisetum pedicelletum	+	0,2	fe3fr3	Poaceae
Spermacoce stachydea	+	0,4	fe3fr3	Rubiaceae
Ampleliocissus leonensis				
Smultis triata	+	0,1	fe2	Vitaceae
Justicia insularis	+	0,2	en1	Convolvulaceae
Angleratum gracillimum	+	0,1	en2	Lamiaceae
Desmodium torto	+	0,4	en1	Fabaceae
Nelsonia canescens	+	0,3	fl	Acanthacea
Schizachvrium nodulosum	+	0,5	en2	Poaceae
Cymbopogon giganteus	1	2, 3	fr4	Poaceae
Chommelina benghalensis	+	*0 ,7*	*en1*	*Commelinaceae*
Paspalum conjufera	+	*0,9*	*en2*	*Poaceae*
Rottboellia exaltata	*1*	*0,8*	*en2*	*Poaceae*
Monechma ciliciatum	+	*0,1*	*en1*	*Acanthaceae*
Hibiscus asper	+	*0,3*	*en2*	*Malvaceae*
Scoparia dulcis	+	*0,4*	*en2*	*Plantaginaceae*
Chloris pilosa	+	*0,9*	*en2*	*Poaceae*
Dicoma tomentosa	+	*0,4*	*fr3*	*Asteraceae*
Spermacoceae chaetocephala	+	*0,1*	*en2*	*Rubiaceae*
Sida ovata	+	*0,3*	*fe2*	*Malvaceae*
Cyperus sp	+	*0,3*	*fe3*	*Cyperaceae*
Cyperus compressus	+	*0,2*	*en2*	*Cyperaceae*
Sporobolus festivus	*1*	*0,6*	*fr3*	*Poaceae*
Justicia kotschyi	+	*0,3*	*en2*	*Convolvulaceae*
Ctenium elegans	*1*	*0,8*	*fr3*	*Poaceae*
Alysacarpus ovalifolus	+	*0,4*	*en2*	*Fabaceae*
Schoenefeldia gracillis	*2*	*0 ,9*	*fr3*	*Poaceae*
Chloris pilosa	+	*0,4*	*en1*	*Poaceae*
Digitaria cilaris	+	*0,5*	*fr4*	*Poaceae*

Eragrostis tremula	*3*	*0,7*	*fr3*	*Poaceae*
Aristida funiculata	*1*	*0,5*	*en2*	*Poaceae*
Vernonia perrottetii	+	*0,4*	*en1*	*Asteraceae*
Setaria pumula	+	*0,8*	*fr3*	*Poaceae*
Ludwigia erecta	+	*0,3*	*fl4*	*Onagraceae*
Euphorbia convolvuloides	+	*0,2*	*en1*	*Euphorbiaceae*
Polycarpea corymbosa	+	*0,3*	*en2*	*Caryophyllaceae*
Stylosanthes erecta	2	*0,2*	*fr3*	*Fabaceae*
Vernonia perrottii	+	*0,3*	*fe3*	*Asteraceae*
Corchorus aestuans	+	*0,1*	*fe4*	*Tiliaceae*
Sida alba	+	*0,2*	*fe2*	*Malvaceae*
Cassia mimosoides	+	*0,3*	*fr3*	*Caesalpiniaceae*
Mitracarpus villosus	+	*0,4*	*fr4*	*Rubiaceae*
Spermacoce stachidea	+	*0,3*	*fe1*	*Rubiaceae*
Allium cepa	+	*0,5*	*fr3*	*Alliaceae*
Bassia scoparia	+	*0,6*	*fefl2*	*Plantaginaceae*
Corallocarpus epigaeus	+	*0,2*	*fl2*	*Cucurbitaceae*
Indigofera vilosa	+	*0,3*	*en1*	*Fabaceae*
Echinochloa colona	*3*	*0,7*	*en2*	*Poaceae*
Aristida funiculata	*1*	*0,9*	*fr3*	*Poaceae*
Monecma ciliatum	+	*0,5*	*fe3*	*Acanthaceae*
Mitracarpus villosus	*1*	*0 ,4*	*en2*	*Rubiaceae*
Vetivera zizanoides	+	*0,7*	*fr4*	*Panicoidea*
Hyptis suaveolens	+	*0,2*	*fe3fr3*	*Labiaceae*
Hydrophila Auriculata	+	*0,5*	*en1*	*Acanthaceae*
Euphorbia aegyptica	+	*0,4*	*en2*	*Euphorbiaceae*
Aspitates ochrearia	+	*0,4*	*en5*	*Geometridae*
Leucas martinicensis	+	*0,3*	*fr3*	*Lamiaceae*
Aristida stipoides	+	*1*	*fr4*	*Poaceae*

Loudetia togoensis	+	*0,9*	*fr3*	*Poaceae*
Achyrantes aspera	+	*0,4*	*fe4*	*Amaranthaceae*
Polycarpea linearifolia	+	*0,3*	*fe3*	*Caryophyllaceae*
Cenchrus bifloris	+	*1,2*	*fl3fr1*	*Poaceae*
Digitaria ciliaris	+	*0,4*	*fr4*	*Poaceae*
Aechelechloa granularis	*1*	*0,7*	*fr3*	*Poaceae*
Sporobolus pyramidae	*2*	*1,4*	*fr4*	*Poaceae*

NB: Fe = foliage fl = flowering Fr= Fructification

Table 2: Herbaceous species eaten on the outskirts of the PNSO

Pennisetum pedicelalatumTrin	+++
Adropogon gayaus Kunth	+++
Panicum maximum	+++
Hyparrhenia ruphens	++
Panicum pansum	+++
Paspalum scrobiculatum	+++
Eragrostis gangetica	+++
Eragrostis pilose	+++
Ipomea coptica	++
Indigofera senegalensis	++
Desmodim hirtum	+
Spermacocoea stachidea	+

Table 2: Herbaceous species. Source: Tchingonta Robert 2023

+++More popular species

++Species eaten moderately, +: poorly eaten

Table 3: Woody species eaten on the outskirts of the PNSO

Species abundance/dominance	Palatability
Acacia Sieberiana	+++
Afzelia africana	+++
Combretum collinum	+++
Daniella Oliveri	+++
Ficus capensis	++
Isoberlinia doka	+++
Kaya senegalensis	+++
Maerua angolensis	+++
Piliostigma reticulata	++
Stereospermum kunthianum	+++
Ziziphus Mauritiana	+++

Table 4: Woody species eaten Source: Patehalet Zouayané, March 2022.

28- Transhumance, a pressure factor on vegetation

Transhumance is the mass movement of livestock in search of better pasture and water. In Chad, there are several types of livestock movement, although the major characteristic of livestock farming is transhumance (Béchir, 2010). In search of forage and water, transhumant herders use a variety of pastoral practices, based on mobility and the adaptation of daily rhythms to the availability of food and water. In this peripheral zone of the Sena-Oura National Park, transhumants stay from October to April. During their mobility, these nomads severely prune woody plants for livestock feed.

Herbaceous primary production was assessed using the phytosociological inventory method and the herbaceous biomass of species. The method used is that of integral harvesting (Levang and Grouzis, 1980). It involves harvesting all plant material (flush cutting) using destructive shears in the 120 plots carried out for the 30 plots installed around the PNSO.

For this purpose, a $1m^2$ square metal frame was placed in 4 locations, so as to integrate the heterogeneity of the terrain. The herbaceous plants present were mowed to 10 cm from the ground, the sample was weighed, then placed in an oven at 103°C for 48 hours in order to assess the dry weight (Boudet, 1981, Honvou et *al.,* 2018). After drying, this amount of biomass was measured and applied to a recovery rate to obtain the amount of biomass on the entire surface.

Overgrazing, one of the factors exerting pressure on plant cover The periphery of the PNSO is home to a variety of pastoral resources. This zone is coveted by breeders and farmers in all directions. It should be noted that in the rainy season, the relics of the par are the only places of refuge for livestock. The herds of sedentary farmers are driven there daily, returning to the village in the evening. The herds of sedentary non-native herders (Foulbés and Arabs) stay for a few months (June-October), i.e. until the end of the rains. At the end of the rains, however, the post-cultivation rangelands are gradually replaced by woody pastures. Indeed, the availability of pasture in the area leads many herders to park their flocks in the same saturated area for long periods. In this terroir, 10326 head of livestock (cattle, sheep and goats) are parked in the same area, creating the risk of severe overgrazing.

29- Trampling and grazing by animals

(Ntoupka, 1999) and (Yonkeu, 1993) cited by Mongo and Ngouné (2015) demonstrate that, if the livestock load is reasonable, grazing promotes wood production by considerably reducing the vigour of flames on tree crowns. Thus, livestock farming can be a factor in desertification if herd load is not controlled.

Through trampling and grazing, regeneration is compromised. The trampling of these animals has an effect on herbaceous and woody plants, as their hooves prevent the plant cover from regenerating. This leaves the ground bare of grasses and plants.

30- -Bush fires, destructive use of vegetation cover

The use of bush fires is a simple and less costly practice. Breeders use it to obtain fodder resources. According to these herders, bush fires encourage the regrowth of young plants. (Boutrais, 1983), believes that bushfire is a normal and indispensable ecological factor for the maintenance of Sudanian savannahs. (Grandi, 1996) quoted by Mongo and Ngouné (2015) believes that fires generally leave bare soil, with no vegetation cover in the Sudano-Sahelian zone. However, it should be noted that pastoral practices based on the use of bushfire are one of the factors that destroy vegetation cover. The use of fire also enables herders to obtain fodder resources.

- Simple linear regression model

This model uses the integral or maximum of the DMP (x) of the vegetation growing season as the independent variable (or explanatory variable) to estimate the above-ground phytomass measured in situ (Y, dependent variable or explained variable) $y = ax+b$.

The method used here to assess annual phytomass originates from work carried out in Niger and Senegal (Wylie et *al.*, 1991; Diallo et al., 1991). It comprises two stages: The measurement of total aerial mass in situ (sites in pastoral zones), using the flush cutting technique, the sampling of herbaceous plants from 30 square yields, and the processing of earth observation data extracted by DMP from the series of images from LANDSAT (Satellite for Observation of Vegetation on Earth) during the vegetative growth period.

32- Analysis of the impact of environmental degradation factors

Household surveys of agropastoralists show in figure 2 that the most damaging factors for the environment are field expansion, development techniques, overgrazing and bush fires.

DAMAGE FACTOR

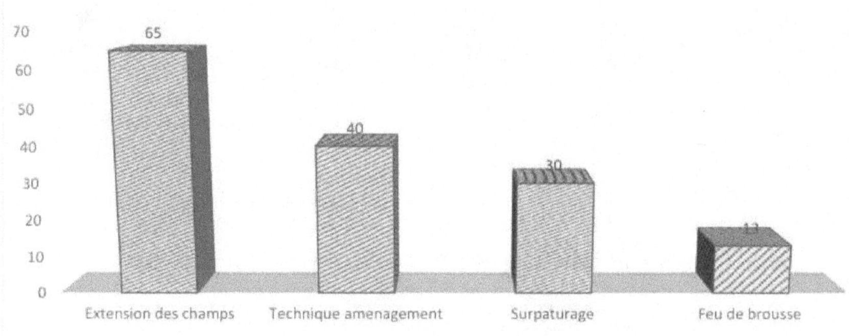

Figure 6: Environmental degradation factor

Figure 2 shows the result of the pressure of agro-pastoral activities on the grazing area, surveyed in 148 sampled households, with 43.91% occupying the agricultural frontage, 27.02% using field management techniques, 20.27% overgrazing and 08.78% bushfire, sampled in 65, 40, 30 and 13 households respectively.

Impact of land use as a factor of environmental degradation

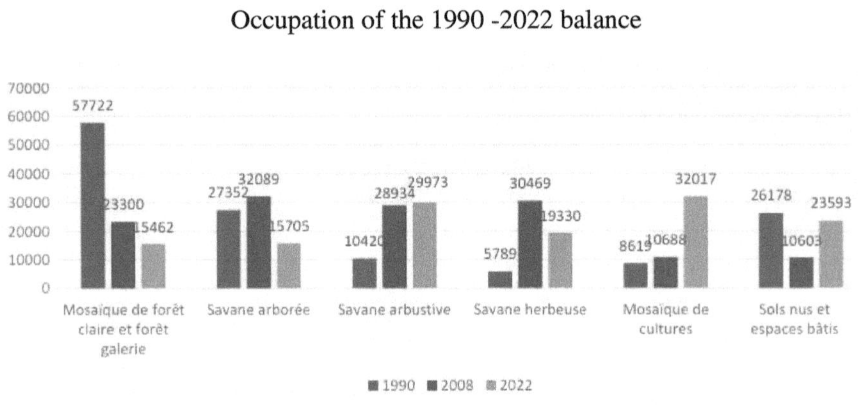

Figure 7: impact of land use as a factor in environmental degradation

The regression of grassy savannah in peripheral areas in favor of agricultural land, observed in the 1990 and 2008 images compared with the 2022 image, can be explained by three main reasons: galloping demography, climate and the increase in the number of livestock in the area. The respective hectare areas in 1990, 2008 and 2022 were 8619, 10688 and 32017. According to statistical data from the Recensement Général de la Population et de l'Habitat (RGPH) in Chad from 2009 to 2022, the population has grown considerably. While the national demographic growth rate is 3.5%. High indeed, but for a total population of the two cantons of 15,000 in 2009, it reaches 21,300 at the 2022 cross rate, i.e. an increase of 70.42% in 14 years.

This study enabled us to assess the pressure of agro-pastoral activities on the grazing area on the periphery of the Sena Oura National Park. The results obtained from our research gave a land use of 1990-2022 respectively at 10688 and 32017 hectares against the total area of periphery of the PNSO of 135,198 hectares. The same findings were observed by the Projet Agri-Elevage de Duras (DCG2-50): Final Report 2007, of the 17,000 hectares covered by the Kourouma terroir in Guinea, the total cultivated area amounts to 4,597 ha. At a rate of 2 ha/UBT/year,

Kourouma's livestock require a grazing area of 11,652 ha. The sum of cultivated and grazed area requirements therefore amounts to 16,249 ha, just about the size of the village. In the commune of Tamou, there has been a rapid increase in the area farmed for this purpose. In thirty years, cultivated areas have increased from 70,764 ha in 1975, or 24.92% of the commune's territory, to 106,111 ha in 2006, or 37.37% of the commune's total area (Moussa, 2014). (Djangrang,2011) in his thesis on "*Pratiques agropastorales endogènes et territorialisation dans la plaine de Mayo-Boneye au Tchad: Etat des lieux et modélisations (1986-2025)* analyzes endogenous agropastoral dynamics and territorialization in the Mayo-Boneye plain in Chad and traces the major stages in the transformations of agropastoral farming patterns, and by analyzing the reciprocal relationships between agrarian dynamics and territorial dynamics.

The number of livestock on the outskirts of the park is galloping, rising from 7501 head in 2012 to 10326 head in 2022, given the growing number of head raised by farmers and the massive arrival of transhumants in the attractive locality. Without relying too much on the discourse that attributes responsibility for the degradation of ecosystems to livestock farming, (Hervé, 1998), consulted in 2010, noted overgrazing in the riparian zone of the Parc du W, with some 770,089 head counted.

Demographic growth is leading to a massive increase in cropping pressure, to the detriment of rangelands, water access areas, grazing areas and transit corridors. This dynamic is reinforced by market demand and the development of entrepreneurial agriculture and agro-business (Colloque régional, N'Djaména, November 23-26, 2021). Securing agro-pastoral land tenure and preventing agro-pastoral conflicts in Central and West Africa.

In the light of the above, we can agree with Guillaud (1993) that "the foreseeable extension of cultivated areas, and due to demographic growth and the current technical context of agriculture, will have the effect both of reducing forage resources themselves and of increasingly limiting access to them".

Conclusion

This study enabled us to analyze the pressure of agro-pastoral activities on vegetation cover. These practices have a negative impact on both the herbaceous and woody strata around the Sena-Oura National Park. The expansion of agricultural areas around the park is linked to the development of cash and food crops. The increase in livestock numbers in this area is leading to a search for pastoral resources, which is having a negative impact. Anthropogenic activities (agricultural and pastoral) appear to be the main factors putting pressure on the vegetation around this protected area. During surveys, the factors leading to pressure on the vegetation cover were identified as population growth, widespread poverty, the creation of new villages in search of fertile land, overgrazing and transhumance. These factors have a major impact on vegetation pressure. With the pressure of rural populations on natural resources, sustainable management actions are essential.

References and bibliographies

Abdelkérim Youssouf, B., Korkorna Moussa, B., (2016). *Effets de l'exploitation de l'or sur la végétation ligneuse dans le Mayo-Kébbi Ouest: Cas de Gamboké et de Yapala,* Mémoire de DIPES II de Géographie, ENS de Maroua, 117p.

Abdoulaye Wouyak, H., Bayi Mayo, A-E., Saidou, K., 2015. *Evaluation de la dégradation des ressources ligneuses sous les effets des pratiques agropastorales en zone périphérique du Parc National de Mozogo-Gokoro (Extrême Nord-Cameroun),* Mémoire de DIPES II de Géographie, ENS de Maroua, 131p.

Akpo LE, Banoin M, Grouzis M. 2003. Effect of trees on forage production and quality of herbaceous vegetation: pastoral assessment in a Sahelian environment. Revue Élev. Méd. vét. Pays trop. 154 (10): 619- 628.

Ali Brahim B., 2010. *Productivité, dynamique des parcours et pratiques d'élevage bovin en zone soudanienne du Tchad,* PhD thesis in Animal Productions, Université Polytechnique de Bobo Dioullasso, 358p.

Boutna, A., 2012. *Dynamique des systèmes agraires et mode de gestion de l'espace. Etude appliquée au Bassin versant du Mayo-Kébbi (Sud-Ouest du Tchad),* Doctoral thesis in Geography, University of Lomé (Togo), 481p.

Boutrais, J., 1983. L'élevage soudanien des parcours des savanes aux ranchs, ORSTOM, 148p.

Regional symposium, N'Djaména, November 23-26, 2021. Securing agro-pastoral land tenure and preventing agro-pastoral conflicts in Central and West Africa.

Djangrang, M., 2011. *Pratiques agropastorales endogènes et territorialisation dans la plaine de Mayo-Boneye au Tchad: Etat des lieux et modélisations (19862025), PhD Thesis in* Geography, University of N'Gaoundéré, 433p.

Guillaud D. 1993. L'ombre du mil: Un système agropastoral en Aribinda (Burkina Faso), ORSTOM, Paris. 321 p

Herve, D., 1998. Animal carrying capacity as an indicator of pressure on forage resources, [online[URL http://

www.documentation.ird.fr/hor/fdi:010013720. Accessed November 10, 2010, 13p

Mamday, 2016. Sampling and power calculations for an impact evaluation, 39p

Mongo, J-C. and Ngouné Sonfack, G-A., 2015. *Dégradation des ressources ligneuses dans les zones périurbaines de Mokolo,* DIPES II de Géographie dissertation, ENS de Maroua, 124p.

Moussa M.S., 2014. Pressure indicators for sustainable natural resource management on the periphery of the W du Niger biosphere reserve. Unique doctoral thesis in geography Université Abdou Moumouni, Niamey, 258p.

Philipeau G., 1986. How to interpret the results of a principal component analysis? STAT-ITCF, 63 p.

Sama O. 2003. Bébédjia, d'une zone de cohabitation pacifique à un espace disputé au sud du Tchad, Master's thesis, University of Ngaoundéré, 118 p.

Sougnabé, P., 2010. *Pastoralism in search of space in the Chadian savannah. Des Peulhs autour de la forêt classée de Yamba-Berté,* PhD thesis in Socio-economics of *development, Ecoles des Hautes Etudes en Sciences Sociales de Paris, 419p.*

Wafo Tabopda, G., 2008. *Protected areas in the Far North of Cameroon: between conservation policies and local practices.* Doctoral thesis in Geography-Planning-Environment, Université d'Orléans.

FSC
www.fsc.org
MIX
Papier | Fördert gute Waldnutzung
FSC® C083411

Kolibri 360 GmbH
Ferdinand-Jühlke-Straße 7
99095 Erfurt, Deutschland
produktsicherheit@kolibri360.de

Druck:
CPI Druckdienstleistungen GmbH
im Auftrag der
Kolibri 360 GmbH
Ferdinand-Jühlke-Str. 7
99095 Erfurt